The Log of the Whaler Helena

Whaler "Helena" of Sag Harbor in the South Pacific 1843-1845

---

by Claus Hoie, able-bodied seaman steersman, and navigator

First edition published in 1994 by

TWO BYTES PUBLISHING, LTD.

219 LONG NECK POINT ROAD

DARIEN, CONNECTICUT 06820

ISBN: 1-881907-16-3 [SOFTCOVER EDITION]
ISBN: 1-881907-15-5 [HARDCOVER EDITION]
ISBN: 1-881907-08-2 [LIMITED EDITION]

TYPESET, PRINTED AND BOUND BY THE STINEHOUR PRESS, LUNENBURG, VT
THIS IS THE FIRST BOOK TO BE SET IN THE DIGITIZED VERSION OF THE
"FELL" TYPES CREATED BY JONATHAN HOEFFLER FROM THE 17TH CENTURY
FONTS IMPORTED FROM HOLLAND FOR THE OXFORD UNIVERSITY PRESS.
DESIGNED BY JERRY KELLY

PUBLISHED AND PRINTED IN THE UNITED STATES OF AMERICA
1994

PHILIP APPLEMAN, Distinguished Professor of English at Indiana University, has published six volumes of poetry, the latest of which is *Let There Be Light* [HarperCollins, 1991], three novels, and half a dozen nonfiction books, including the Norton Critical Edition, *Darwin*. His poetry has won a fellowship from the National Endowment for the Arts, the Castagnola Award of the Poetry Society of America, and the Humanist Arts Award of the American Humanist Association.

A former merchant seaman, he now lives in Sagaponack, Long Island — sister village of the old whaling port, Sag Harbor — with his wife, the playwright Marjorie Appleman.

Foreword BY PHILIP APPLEMAN

SINCE LIFE ON THIS PLANET probably began in some kind of primordial soup, and since that nourishing broth may well have been the ocean, it's not surprising that so many of us feel the sea "in our bones"—or perhaps in our blood, which is as salty as the ocean itself. Tightly knit up with that affinity for the sea is our long fascination with the ships that sail on it. Since their first known appearance, eight thousand years ago in Egypt, a multitude of variants on that old design have shown us their seductive profiles: Greek argosies, Roman triremes, long, single-masted Viking raiders, rangy caravels, great, heavy-canvassed galleons, slender-hulled frigates, moon-raking clippers, sleek schooners—and of course, square-rigged whalers like Claus Hoie's fictive bark, the *Helena* (affectionately named for Claus' wife, Helen, who is a superb artist in her own right).

In Norway, as in other seafaring nations—England, Greece, Portugal—nobody is ever very far from the sea, and a substantial proportion of young men become sailors as a matter of course. Claus Hoie, Norwegian by birth, went to sea as his father and grandfather did, and now, as coincidence would have it, lives near one of the great nineteenth-century whaling ports, Sag Harbor, where he has had the opportunity to study the ships' logs of those old whalers, Captain Ahab's contemporaries.

But the captains of the nineteenth-century whalers were not obsessed like Ahab. Part seaman and part entrepreneur, they not only had long years of experience behind them, but they also got a large share of the profit when their ships came in; so their dutiful logs convey both the wild adventure and the hard-nosed business of the sea. Laconic, precise, matter-of-fact, they record financial profits and personal tragedies over the year, or two years, or several years of those epic voyages. Claus Hoie's typically succinct chronicle of the *Helena* gives us a glimpse into the details of that life: the seamen's daily round of endless maintenance, repairing, painting, splicing, coopering, blacksmithing; the labor of a hard chase; the task of transforming their gigantic prey into thousands of barrels of whale oil and tons of whalebone; the scrimshaw boredom of the seaman's off-watch hours—all of that in the periods of relative calm between the ferocious storms that could tear away their sails and bring on the dreaded cry, "Man overboard!"

Seagoing writers have interpreted all this for us as seamen themselves never would. (My own shipmates, a generation ago, tended to view the ocean chiefly as a spiteful inconvenience for ships.) Conrad gave us the romance and tragedy of seagoing; Melville the intriguing conjunction of the humdrum and the mysterious. But a painting can show us even better than words the light and shadow, the gaunt visage, the sheer muscle of it all—and even, in wild artistic surmise, under a yellow Hoie sky, a seaman's fantasies and dreams. Claus Hoie's fine work does us all that service.

There were no save-the-whale movements in the nineteenth century; they were not yet necessary because the efficient, impersonal, and murderous twentieth-century factory ship had not yet been invented. Even when the traditional whalers, the gallant Queequegs of the South Pacific, giving chase in small boats, with hand-held harpoons, made a profitable voyage, their greatest successes were not sufficient to endanger a cetacean species. The reason we, and Claus Hoie, continue to marvel at those old larger-than-life harpooners is that they were pitting themselves, one-on-one, against Goliath. ("Larboard boat struck," writes Hoie's Captain Edwards, "and he stove the boat.")

We are lucky that Claus Hoie's rich experience as seaman, as artist, and as connoisseur of the old ships' logs has turned out to be an irresistible artistic challenge for him: this beautiful edition, with its haunting portraits of a lost way of life, is our resulting good fortune.

Barque "Helena" Captain John C. Edwards departs Sag Harbor on a whaling voyage to the South Pacific and elsewhere on February 3d 1843

Capt. John C. Edwards
Claus Hoie

Saturday April 2nd 1843

This day commences with pleasant weather the crew employed fitting our boats fore whaling and we chose the boat crews. Middle and latter parts fresh breezes from the Northward steering East by S. Some rain and squally took in fore and main top gallant sails. At 5 saw a sail on our weather bow standing to the Westward

So ends these 24 hours

Lat by obs 26.13 N
Long. 58.23 W

N

Wednesday July 25th 1843

Commences with light winds. Middle heavy winds from the Westward. At 10 AM took in the top gallant sails and flying jib. At 6 PM the winds abated and we made sail steering ESE. Four ships in sight one of which crossed our stern bound for the Northward With a long voyage ahead we wonder about our fate — to be so long and so far from home and loved ones we pray for heavenly protection from the stormy seas and the monsters of the deep. So ends this day

Lat. 18.20 S
Long. 46.13 W

Monday Aug. 16th 1843

Commences with light breezes from the S.W.
saw whales lowered and chased Larboard boat
struck but lost their whale three miles
from the ship. Middle and latter parts the
same. A ship passed to the Windward of
us. Saw whales and chased to dark
without success.

So ends these twenty-four hours

Lat 36.23 Long 15.18 W

Tuesday Nov 25th 1843

Begins with light winds from the Northward
saw plenty of whales lowered all boats
and chased Starboard boat struck and

drawed. Waist boat and larboard boat
chased without success. Middle part fresh
breezes from the N.W. Saw sperm whales
and chased. starboard boat struck 
S.B.
and at six o'clock brought the whale
alongside to commence cutting in on
the morrow. So ends these twenty four hours

Lat. 22.13 N
Long. 55.21 W

Thursday Jan. 12th 1844

This day begins with light breezes
from the Southward Saw several sails
and many whales lowered and chased 
Starboard boat struck but he stove
the boat and the boatsteerer managed 
to cut the line and let him go. Larboard
boat struck and drawed, one man fell from
the boat but was pulled in All hands safe.
So ends these twenty-four hours

Lat. 12.32 S
Long 164.16 W

Wednesday Feb 23d 1844

Commences with light winds from the West Middle and latter parts the same. at one oclock raised a sperm whale off our starboard bow and Mr. Jones fastened to him and he took his line. The Mate fastened to him about ten miles from the ship and killed him. Since it

S.B.

being calm we could not work the ship to him so the Capt's boat took lines in buckets back to the whale and commenced towing him until we got him alongside about ten at night.

So ends this day Lat. 2.03 N Long. 172.16 E

Thursday March 6th 1844

Comes in with light winds from the Westward steering SE by E At 4 PM saw two right whales lowered and kill one and took him alongside and started cutting at six o'clock finished at ten, stood greater watches will call all hands at four to start the works to commence boiling on the morrow.

W.B.

So ends these 24 hours

Lat. 14.03 N
Long 172.16 E

Tuesday March 19th 1844

Begins with light winds from the Eastward searched for whales without success Middle part fresh breezes and there are heavy clouds on south horizon sign of bad weather ahead. Took in top gallant sails and flying jib. Steering South by W. now heavy rain squalls.

So ends these 24 hours. Lat. by obs. 5.23 N

Long. 170.63 W

Wednesday Mar 20 1844

the storm struck at daylight and it was upon us very fast. Mainsail was torn away and larboard boat was smashed. At noon the mate shouted man overboard but there was nothing to be done in the heavy seas. God have mercy on his soul. so ends this day.

Lat. by est. 9.21 N
Long. 172.31 W

Sunday April 10th 1844 

Have put in to the island of Kuola Pur to repair the damage of the dreadful storm It will be some time before we can return to the cruising grounds. The crew has been given ~~the~~ shore leave. There is much to be done and all hands will be busy with the ship's work, but Seaman Johnson has been missing for three days.

So ends these 24 hours Lat. 8.03 N Long 164.12 W

Thursday July 26th 1844

Comes in with light winds from the Eastward with fog. Spoke a Holland ship did not see her  until she was within a quarter mile the fog was so thick. At 2 P.M. the fog lifted and we raised a whale close lowered and as good luck would have

S.B. it we got him. at 5 oclock we got him cut in and started the works. Saw several finbacks on our Starboard side. So ends these 24 hours.

Lat. by est 21.20 N
Long by est. 157 E

Monday October 23d 1844

Commences with mild breezes from the Southeast. Spoke barque Argus from Nantucket homeward bound after twenty-seven months with 1800 barrels of oil and 15,000 lbs. of bone. We gammed for three hours and gave them letters to our loved ones at home and wished them Godspeed on their voyage. So ends these 24 hours

Lat 27.03 N Long 167.13 E

Tuesday Jan. 26th 1845

Begins with strong breezes from the South
Middle part strong winds from the Southeast.
At noon seaman Smith spoke with insolence
to the captain when he was ordered to go aloft
~~the~~ The Capt. was obliged to have him
punished with three lashes of the cat-o-nine tails
and to order him put in irons where he is
now. Latter part mild breezes from the S.E.
Saw sperm whales and chased without success.
So ends these twenty-four hours

Lat. 34.42 Long 177.31 E

Wednesday April 12th 1845

We have decided to start our homeward voyage. Fresh breezes from the WSW At 7 AM had a very sudden and heavy squall but generally good weather for this part of the world. Calculate we passed Cape Horn this morning. Saw a ships to leeward heading westward. Our hearts are light to be homeward bound.

So ends this day Lat by est. 56.23 S Long 58.16 W

Thursday June 23 1845

Commences with light breezes from the W inclining Northward, hands employed at ship's duty, painting a boat, making spun yarn, life buoys etc. Cooper and Blacksmith at their trade.

Spoke the Lucy Ann of Greenport outward bound. 5 studding sails and 3 royals out which we have carried week in and week out.

Latter part light air and fog.

So ends this day Lat 6.70 N. Long 38.16 W

Tuesday July 16, 1845

This day commences with moderate breezes from the North. Saw a brig steering outward. Hands employed in various jobs. At 3 saw Montauk  Light and gave her two guns. at ebb tide came to anchor in 10 fathoms of water four miles west of Montauk. On the morrow we shall take a pilot and go up the bay to our Sag Harbor destination.

So ends this day

Lat. 41.12 N Long. 71.87 W

Wednesday July 17th 1845

This is the the last day of our journey We arrived safely at Sag Harbor wharf with thankful hearts. Our holds are full with 2,100 barrels of whale oil, 226 barrels of sperm oil, and 16,000 pounds of whale bone. It has been a long journey and we are happy to see our families alive and well.

So ends this voyage.

SAG HARBOR

THIS JOURNEY actually began in Stavanger, Norway, where author/artist Claus Hoie was born. In very early life he came to America with his family. After finishing his schooling he went to sea as most Norwegians would—in the Merchant Marine where he learned to be able-bodied seaman, steersman and navigator.

A great love for ships and the sea has always remained with Claus Hoie, but as his love for art was greater, he returned to school to study art at Pratt Institute in New York and at Ecole des Beaux Arts in Paris. Combining his artistic talent with his love of the sea, Claus Hoie in his book, *The Log of the Whaler Helena,* steps into the shoes of Captain John C. Edwards to tell the story of the dreams and adventures of 19th-century whaling.

During World War II Claus Hoie joined the special Norwegian-American infantry in the United States Army, which were prepared as ski troops in the event of an invasion to liberate Norway. It was while training in this battalion that he took time to paint a number of watercolors of the mountain training they were undergoing. These paintings have been exhibited at the Denver Art Museum, the University of Minnesota and at the Norwegian-American Museum in Decorah, Iowa. Ten years ago the paintings were shown at the Akershus Museum in Oslo, Norway, where they now reside as part of the permanent collection.

Claus Hoie's work shows his great fondness for the sea, that which sails on the sea, and the creatures that live within. Many awards have been bestowed on this distinguished artist including the most recent Thomas M. Hoyne III Award at The International Exhibition at Mystic Seaport Museum in 1993, Award for Painting from the American Academy of Arts and Letters, the National Academy of Design's John Pike Memorial Award, and the Gold Medal of Honor from the American Watercolor Society. Some of his paintings are part of permanent collections of museums including the Brooklyn Museum, South Street Seaport Museum, Oklahoma Museum of Art, Butler Institute, Norfolk Museum, Center College of Kentucky, Guild Hall Museum of East Hampton, and Brigham Young University.

The Log of the Whaler Helena is Claus Hoie's special masterpiece combining the skill of an author for storytelling with that of an artist as a painter depicting the sights and feeling of the sea and as a calligrapher in the penning of this log. This book is indeed a work that author and artist can be proud of.

Come Aboard! Join Captain John C. Edwards on the whaling bark, *Helena,* bound for the South Pacific and elsewhere. See and feel this adventure for yourself through the eyes of an experienced seaman, Claus Hoie.

. . . So begins the journey . . .

TIM LEE